THIS BOOK BELONGS TO:

_____

The Wonderful World of

DRAGONFLIES

Mimi Jones

Dedicated to all who love dragonflies!

ISBN 978-1-958985-12-0

www.joeysavestheday.com

A Mimi Book

Welcome to the Wonderful World of Dragonflies.

Dragonflies are flying insects!

There are over 5,000
species of dragonflies

Some scientists say that dragonflies have been here long before dinosaurs ruled the earth!

The scientific name for dragonflies is Anisoptera.

The globe skimmer, also known as the wandering glider dragonfly is one of the most common types of dragonflies.

RARE

The Hine's emerald dragonfly,
also known as the Ohio emerald,
is one of the rarest types of dragonflies.

The Roseate skimmer.

There are many colors of dragonflies,
but one of the rarest is pink dragonflies.

Life Cycle of a Dragonfly

Egg

Adult

Larva

Dragonflies are carnivores (insectivores).

Carnivore means that they
only eat meat. Dragonflies eat
flying insects and other insects.

A group of dragonflies is called
a cluster or a flight of dragonflies.

Dragonflies are primarily diurnal and prefer to hunt for food during daytime hours. Being diurnal means dragonflies are awake during the daylight hours.

Diurnal

Day

Nocturnal

Night

Here is a list of names of some of the different types of dragonflies:

- Butterfly Dragonfly
- Damers
- Emerald Dragonfly
- Meadowhawks
- Micrathyria
- River Cruiser
- Tigertail Dragonfly
- Skimmer dragonfly
- Spiketail Dragonfly
- Wandering glider

Dragonflies are cold-blooded
and love to be in the sun.
The sun helps them regulate their
body's temperature.

Dragonflies live around streams, lakes, ponds, and other wetlands.

Stream

Lake

Pond

POOP!

IN CASE YOU WERE WONDERING,
YES, DRAGONFLIES POOP!

Have you ever wondered how these
insects manage their waste?
Let's take a dive into the less talked
about yet equally intriguing aspect
of their excretion (POOP) process.

The Digestive System of Dragonflies:

Dragonflies have a complex digestive system that starts with their mouth, designed to capture and consume prey.

After digestion, waste products are processed in the digestive tract.

Dragonflies have a distinctive feature located at the end of their bodies called the rectal chamber. This specialized structure functions as a storage space for waste materials, a unique adaptation not found in humans.

The Excretion Station:

When it's time for a dragonfly to excrete (Poop), it utilizes the rectal chamber as its command center for waste management.

Dragonflies have a unique way of expelling waste. They use their powerful abdominal muscles to efficiently expel waste from their rectal chamber, allowing them to stay clean even while in flight.

A Flight with a Surprise:

It's fascinating to note that dragonflies have the unique ability to release their waste while in flight. So, the next time you find yourself observing these extraordinary creatures gracefully maneuvering through the air, take a moment to appreciate the fact that they are actively engaged in their natural elimination process.

Dragonflies have large compound eyes that are comprised of numerous ommatidia, or simple eyes.

This complex eye structure enables them to have an almost complete 360-degree field of vision, with just a small blind spot located at the back of their heads.

Full Rotation
360°

180°
Half Rotation

Dragonflies can fly at speeds up to 34 miles per hour.

34 mph

There once was a beautiful emerald-green dragonfly who loved to fly from pond to pond in search of food! One day, she couldn't find flying insects to eat at her usual eating spots on the pond. So, she went on an adventure, flying high in the sky till she came to this large lake.

There were flying insects everywhere. Oh, how happy she was! She spent all day eating all the flying insects she could. She was filled with joy, and her belly was filled with food. She flew back to her pond and slept all night.

Dragonflies are present on every continent except Antarctica.

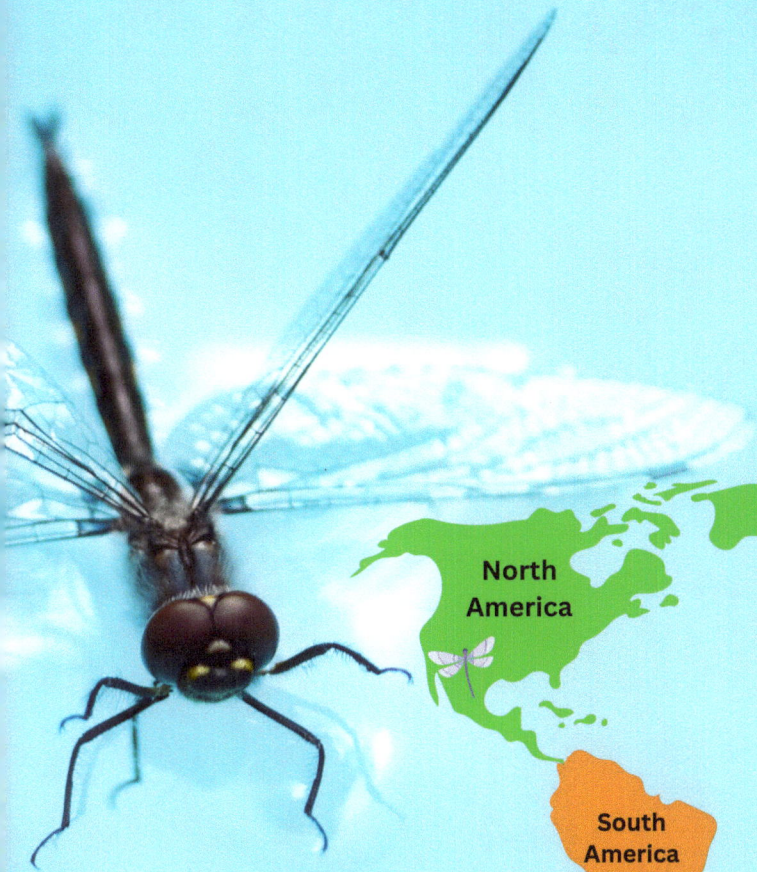

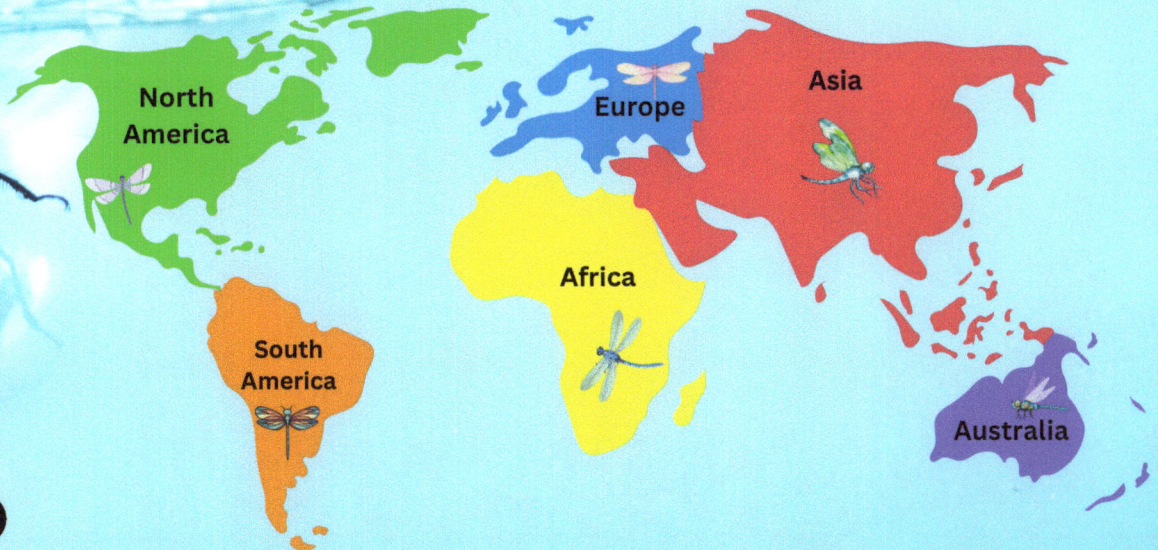

WORLD

North America

South America

Europe

Asia

Africa

Australia

Antarctica

Can you count the dragonflies?

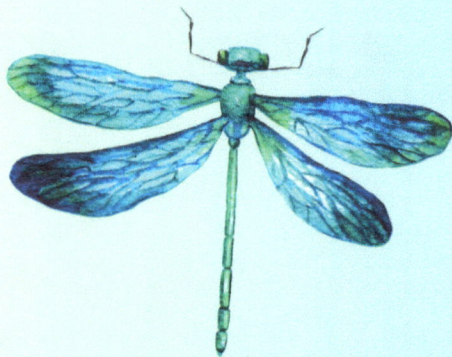

THE END!

www.ingramcontent.com/pod-product-compliance
Lightning Source LLC
Chambersburg PA
CBHW060835270326
41933CB00002B/94